AF396391

THE SECRET HANDBOOK OF THE BLUE CIRCLE

Text: Antonis Papatheodoulou
Illustrations: Iris Samartzi

Concept Development and Scientific Editing: Despo Fatta-Kassinos
© Despo Fatta-Kassinos 2019

Published by

IWA Publishing
Alliance House
12 Caxton Street
London SW1H 0QS, UK
Telephone: +44 (0)20 7654 5500
Fax: +44 (0)20 7654 5555
Email: publications@iwap.co.uk
Web: www.iwapublishing.com

First published 2019
© 2019 IWA Publishing

Apart from any fair dealing for the purposes of research or private study, or criticism or review, as permitted under the UK Copyright, Designs and Patents Act (1998), no part of this publication may be reproduced, stored or transmitted in any form or by any means, without the prior permission in writing of the publisher, or, in the case of photographic reproduction, in accordance with the terms of licenses issued by the Copyright Licensing Agency in the UK, or in accordance with the terms of licenses issued by the appropriate reproduction rights organization outside the UK. Enquiries concerning reproduction outside the terms stated here should be sent to IWA Publishing at the address printed above.

The publisher makes no representation, express or implied, with regard to the accuracy of the information contained in this book and cannot accept any legal responsibility or liability for errors or omissions that may be made.

Disclaimer
The information provided and the opinions given in this publication are not necessarily those of IWA and should not be acted upon without independent consideration and professional advice. IWA and the Editors and Authors will not accept responsibility for any loss or damage suffered by any person acting or refraining from acting upon any material contained in this publication.

British Library Cataloguing in Publication Data
A CIP catalogue record for this book is available from the British Library

ISBN: 9781789061086 (Paperback)

ISBN: 9781789061109 (ePub)

To all the dreams, be they ours or those of others, young and old, which turn into reality because we really believe in them and we try... and we strive for them... and we remain restless... for all those things on which we spend hours, days, years... until they come true... This book is dedicated to all those dreams.

This book was one of my dreams; the dreams of a girl that loved chemistry as a pupil and went on to study Chemical Engineering after finishing school. After years of studying, she became a professor at the University of Cyprus. There, she built an amazing team of brilliant professionals who adore research and continuous learning.

This dream remained locked in my heart, until the efforts of my research team paid off, bringing research funding and success... until the universe conspired in two extraordinary and talented people crossing my path: Antonis and Irida. They immediately understood the wish I described to them and managed to convey everything technical, complicated and maybe generally incomprehensible about 'waterology' to this book you are now holding.

I feel great joy and satisfaction. I hope you all feel the same joy when you grow up and fulfil your dreams. Make sure to put your heart and soul into studying what you love. Because professional progress and career development can fulfil your life and make you truly happy. To quote Dr. Constantinos Christofides, our university's former rector and a Physics Professor, "as soon as we enter the world of work, money is not that important once we have covered our basic needs. What drives people is the satisfaction they get from their job because the game of life is not one we can rewind and replay from scratch". I passionately believe your mission in life is to do what truly makes you happy and to move forward.

I want to thank my colleagues from the bottom of my heart; to all those who helped me open the first Environmental Engineering Laboratory in Cyprus, which grew to become one of our university's research units. It is a big family, which wants to share significant new knowledge regarding the protection of the earth, water and environmental health...

With your contribution this family can grow even bigger. It can grow to become a real community; a community that cares about water. This is how we like to picture every scientist in our sector, as well as all children, parents and teachers, whose awareness we manage to raise through this book. We picture them as the members of a society that takes care of water and protects it so that there is always enough for everybody: of the "Eternal Blue Circle Society".

Antonis, Irida, Irene, Lida, Costas... thank you!

Despo Fatta-Kassinos
Associate Professor
University of Cyprus

Dr. Despo Fatta-Kassinos would like to warmly thank Dr. Norbert Kreuzinger (Vienna University of Technology, Austria) and Professor Lian Lundy (Middlesex University, UK) for their help during the technical English language editing.

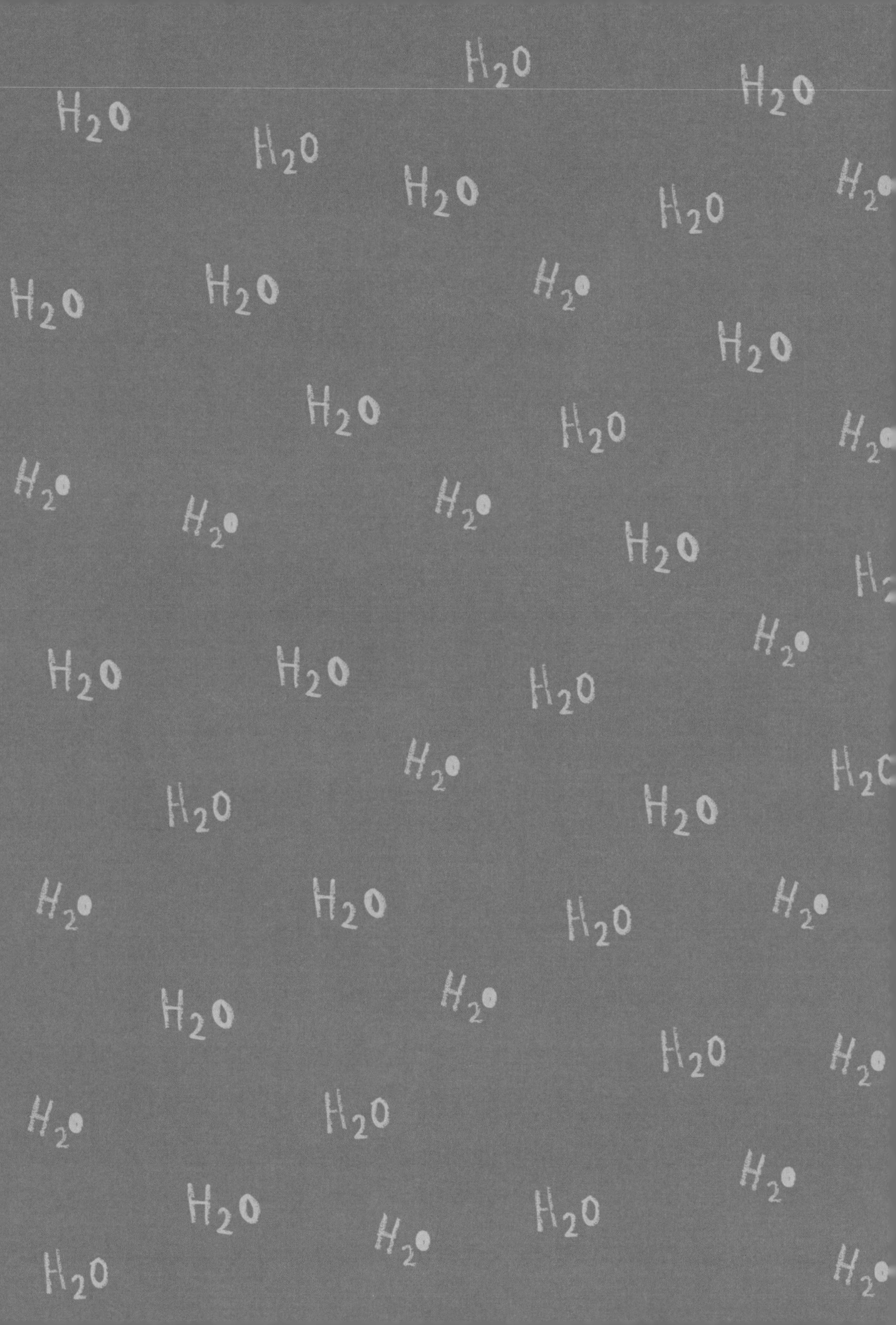

WELCOME

H₂O

Unbelievable! We have visitors!
Dear prospective members! Let me introduce myself:
my name is **Dr. Tom Atom** and I welcome you
to the Eternal Blue Circle Society.
We are the defenders, the researchers, the scientists,
the protectors of the most important substance on the planet: WATER.
And we are just recruiting for helpers – we need you!
In order to help us, you need to know all about us and what we do.
And before anything else, you need to understand what water is.
To understand it well... very, very, very very WELL!

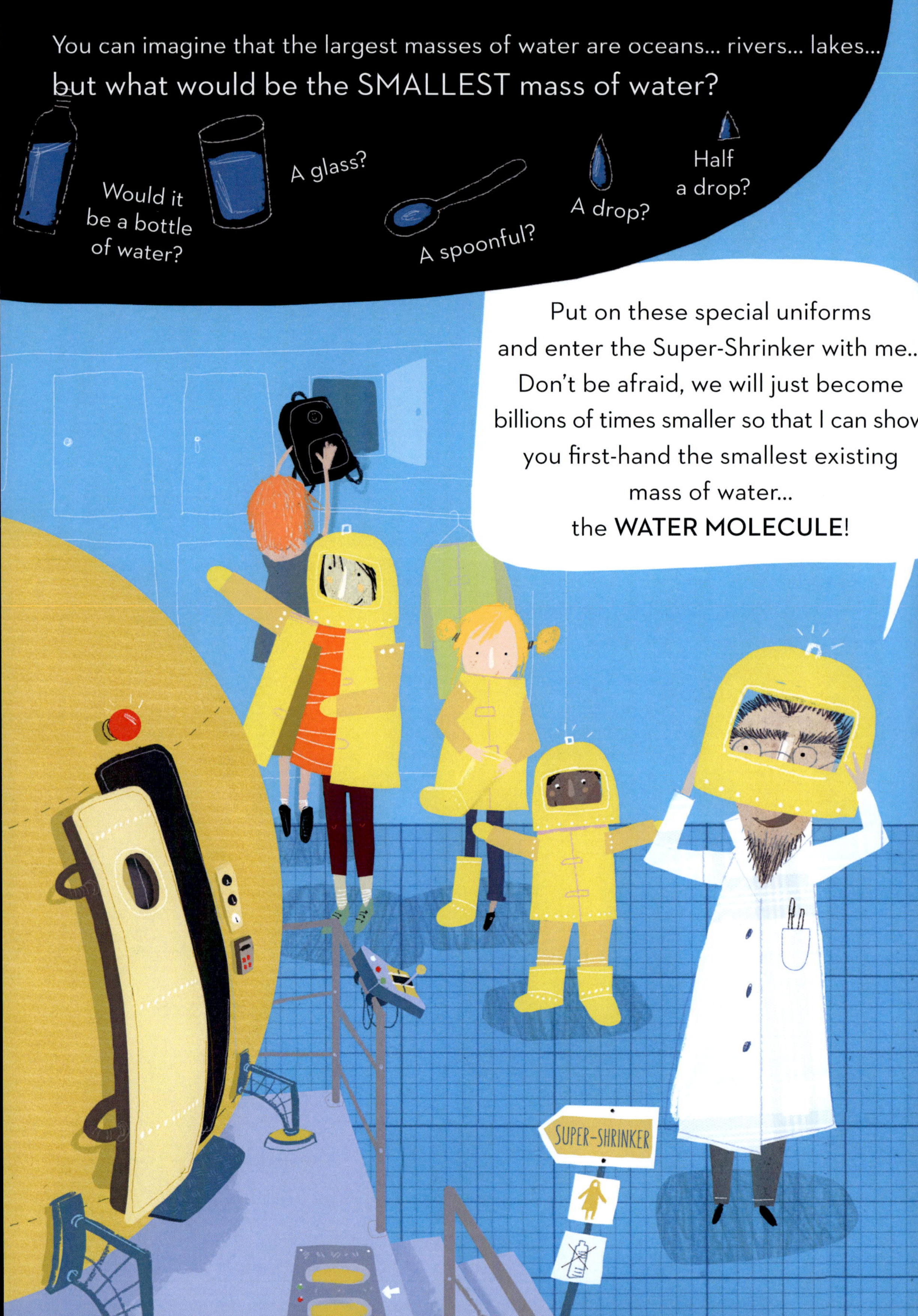

You can imagine that the largest masses of water are oceans... rivers... lakes... but what would be the SMALLEST mass of water?
Would it be a bottle of water?
A glass?
A spoonful?
A drop?
Half a drop?
Put on these special uniforms and enter the Super-Shrinker with me... Don't be afraid, we will just become billions of times smaller so that I can show you first-hand the smallest existing mass of water... the WATER MOLECULE!
SUPER-SHRINKER

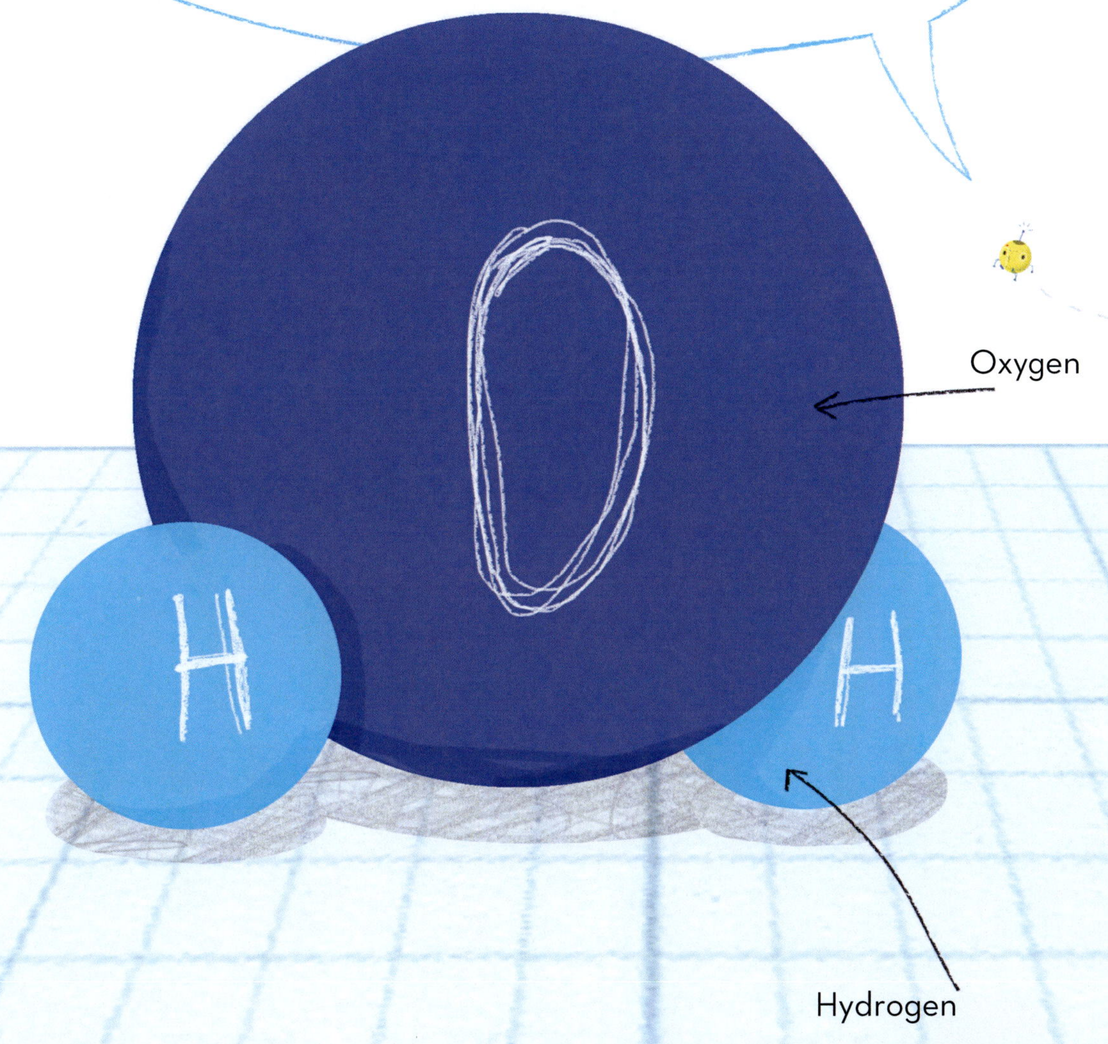

Voilà!
This small group of friends you can see here is in reality
the smallest mass of water.

Three friends that hold on to each other so strongly they
can stay like this for millions of years. One atom of the
chemical element Oxygen (O) holds hands with two atoms
of the chemical element Hydrogen (H). This is why, in
Chemistry, the chemical formula of water is H_2O.

Oxygen
H
O
H
Hydrogen

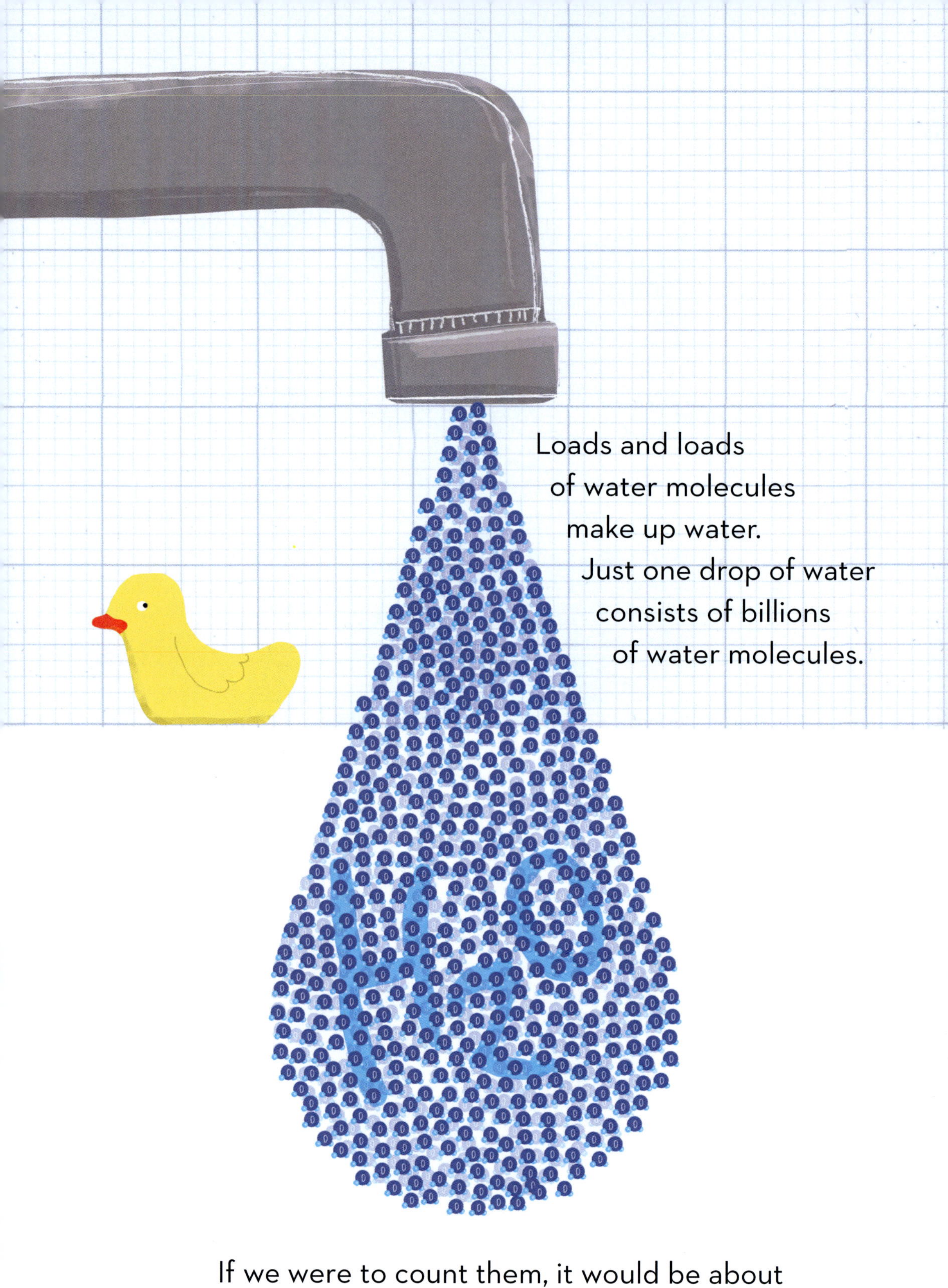

If we were to count them, it would be about
1,700,000,000,000,000,000,000 molecules.

Depending on how its molecules spread
throughout a space, water has different names.
It is called WATER when it is in liquid form,
when it is in solid form, it is called ICE
and when it is in gaseous form
it is called WATER VAPOUR.
WATER
ICE
WATER VAPOUR

When do your toys take up less space in your room?

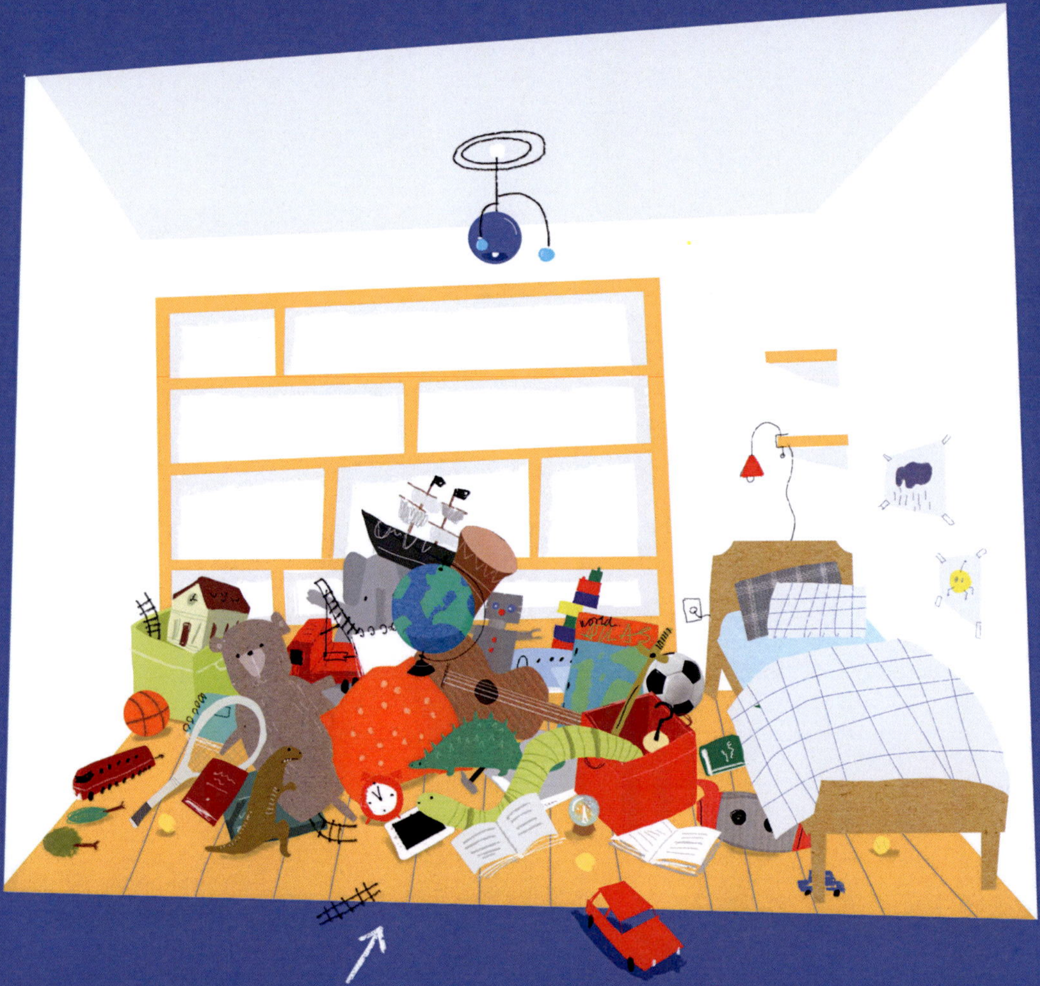

Is it when they are scattered around every corner of the room
or when they are neatly placed on the shelves?

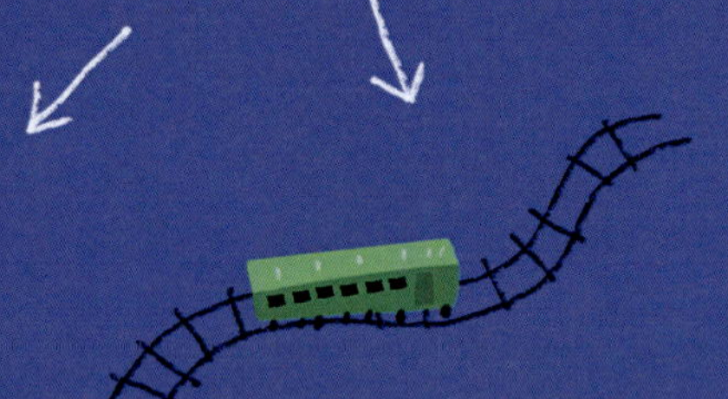

When they are neatly ordered, of course!
Well, the exact opposite happens in the case of magical,
wonderful, strange WATER!

When it is in a solid form, meaning when it is 'tidier',
it takes up more space!
This is why, if we put a bottle of water in the freezer,
as soon as it becomes ice it will not fit in the bottle
anymore and the bottle will break!

Also... water is at the same time the HUMBLEST
and most INVALUABLE substance:
Humble because it has no COLOUR, SMELL or TASTE!
And invaluable because...

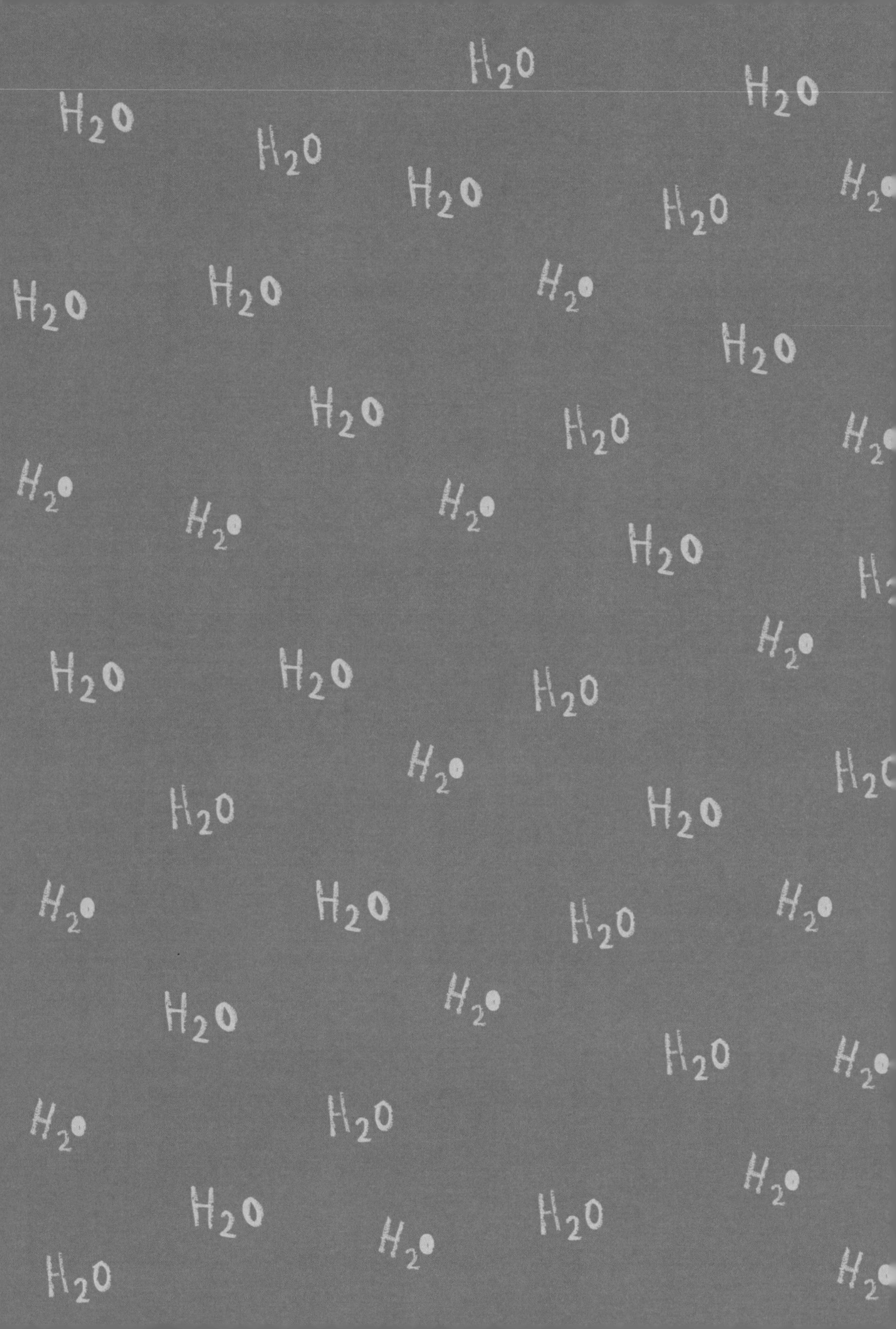

WATER IS EVERYWHERE

in the sky
in oceans
on the land
and lakes and rivers
in the sea
Water takes up the largest part of our planet's surface,
Water is everywhere:
our body, our blood and cells,
our food...

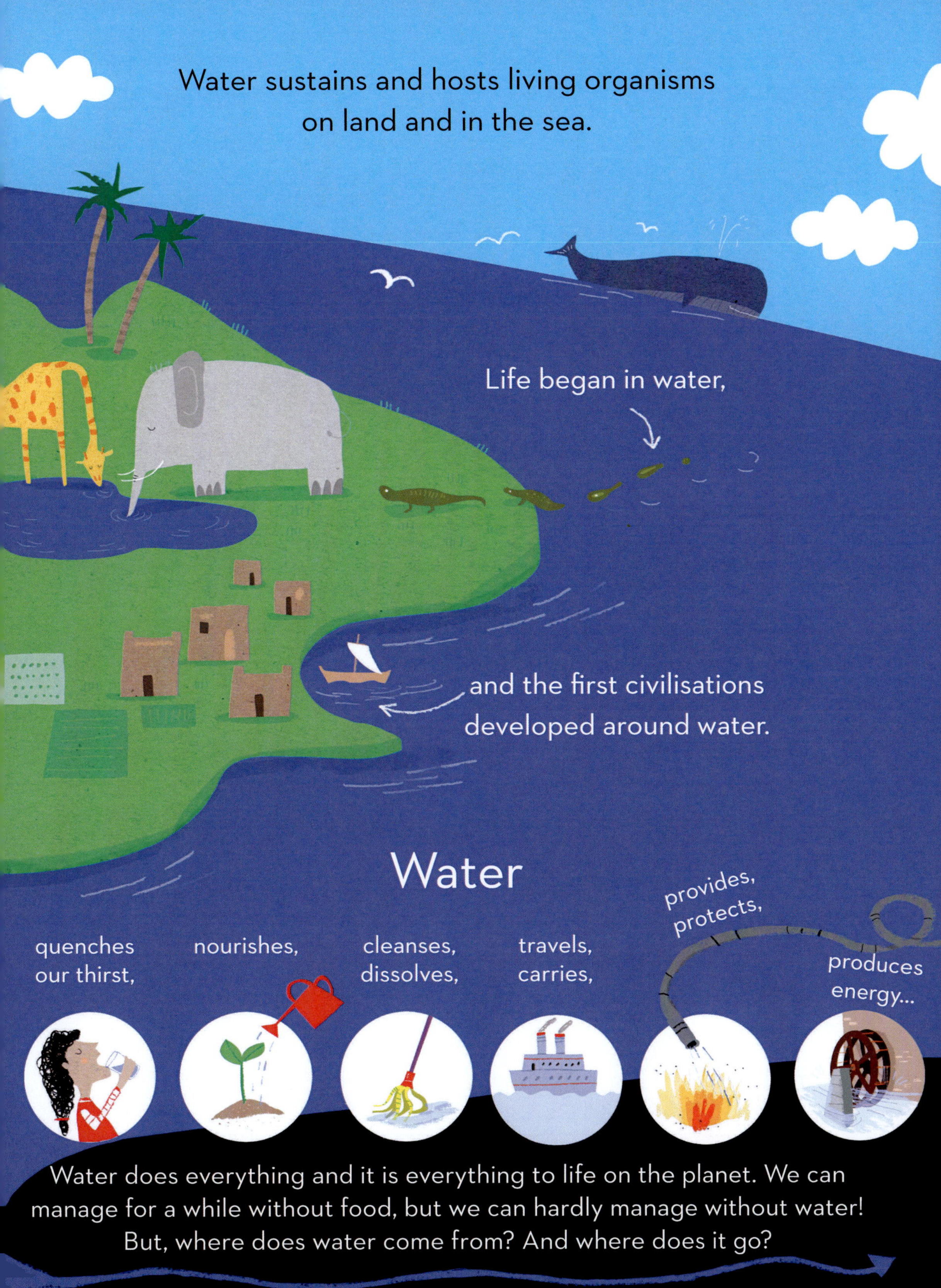

Water sustains and hosts living organisms on land and in the sea.

Life began in water,

and the first civilisations developed around water.

Water

quenches our thirst,

nourishes,

cleanses, dissolves,

travels, carries,

provides, protects,

produces energy...

Water does everything and it is everything to life on the planet. We can manage for a while without food, but we can hardly manage without water! But, where does water come from? And where does it go?

My colleague Dr. Celia Circle will explain that

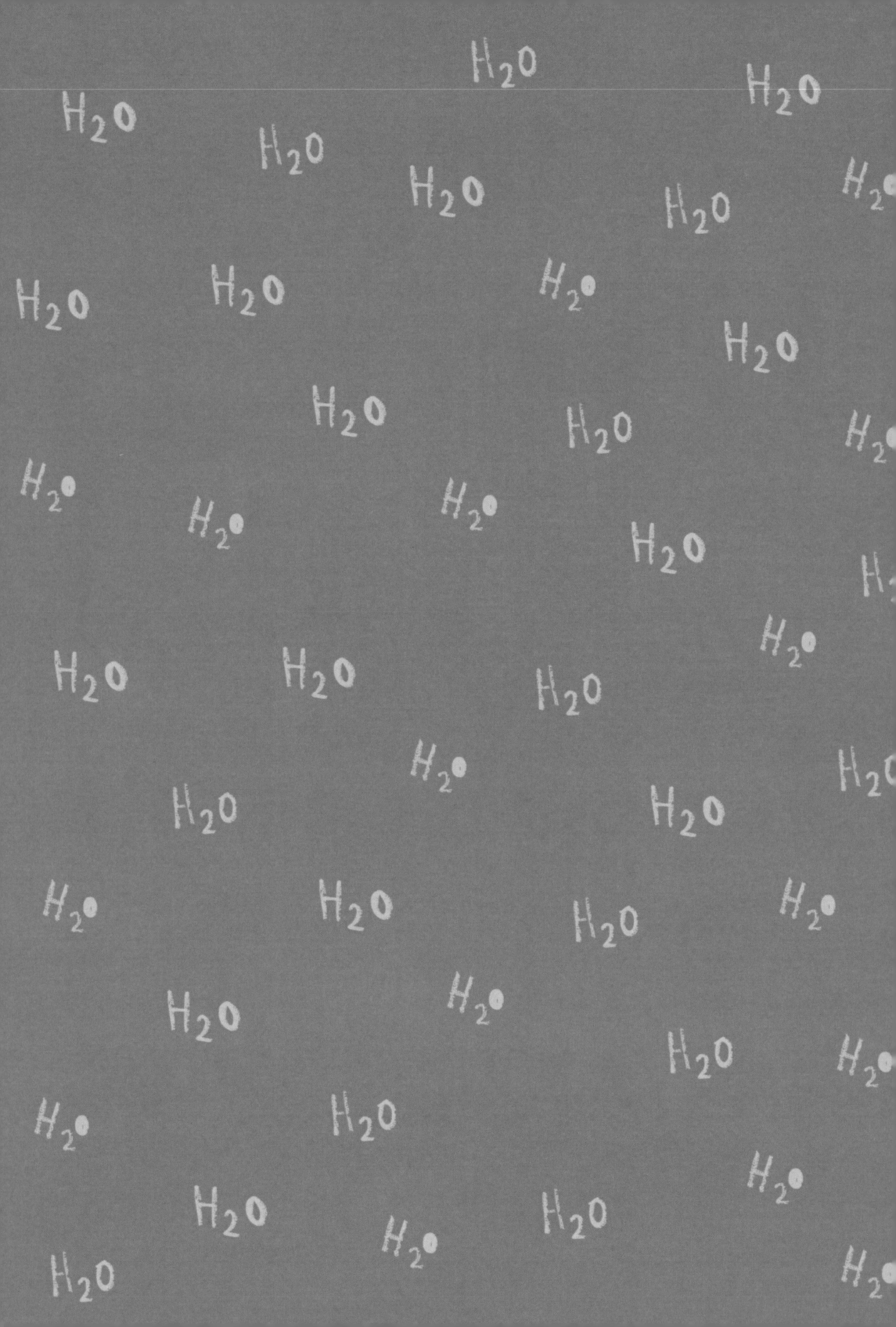

THE WATER CYCLE

Dress warmly because...
we will fly with **Dr. Celia Circle**
to discover water's
journey together.

We do not yet know a lot
about how so much water
appeared on Earth billions
of years ago, but we know
that, since then, the same
water travels continuously
and changes form constantly.

Its journey starts again and repeats itself
where it ends... that is why we call it
THE WATER CYCLE,
an eternal cycle that never stops.

The sun warms water in seas, lakes and rivers. This water vaporises, which means it takes its gaseous form, becoming water vapour and, together with the water vapour coming from growing plants, called 'transpiration', goes up into the sky.

As it goes up in the atmosphere, water vapour gets cold and turns into water drops or even tiny water crystals… which fall back to earth as rain, snow or hail.

Water falls back into the sea or on dry land. Part of the water that falls on dry land goes into the ground and travels to the roots of plants. It is temporarily stored in lakes or deep within rocks. And it ends up back in the sea through streams and rivers…
And there, thanks to the sun, the same journey begins, again.

Can you see this town next to the lake? Let's land and I will tell you all about it, as it also hides other water cycles and routes made not by nature but… by humans!

From ancient times, we have wanted to transport fresh water from springs, lakes and rivers to our towns and fields, to water plants and animals and, of course, to drink it. Ancient people built water supply systems and aqueducts, invented pumps and filters...

Imagine your town's water supply system as a big, well-designed machine that you switch on every time you... turn on the tap!

This water can start out under the ground (in special layered rock formations called aquifers), or a lake or even a reservoir, which is a lake made by engineers who built a big barrier (called a dam) to stop the river's flow.

To kill the bacteria living in the water, they add, for example, chlorine (chlorination) or apply ultraviolet radiation (UV). This stage is called disinfection. Then, water is ready to start its journey through miles of pipes until it reaches our taps, bathroom and kitchen sinks, bathtubs, toilet flushers and boilers in every house in the city!

Water then travels by pipe to a **water treatment plant**.
There, engineers have thought of quite a few tricks to **clean water**,
to make it safe for us to drink, because during the water cycle
many substances decide to travel together with the water.
To get rid of various solid items like sand and grit, they use
sedimentation tanks.

To get rid of very very small solid particles that are in suspension or have
dissolved in the water, they add substances that make solid particles
stick to each other and form bigger, heavier particles known as flocs.
As the water is held in tanks, the flocs sink and, as a result,
they are removed from the water.

They can also use sand filters to clean the water by removing even finer
particles.

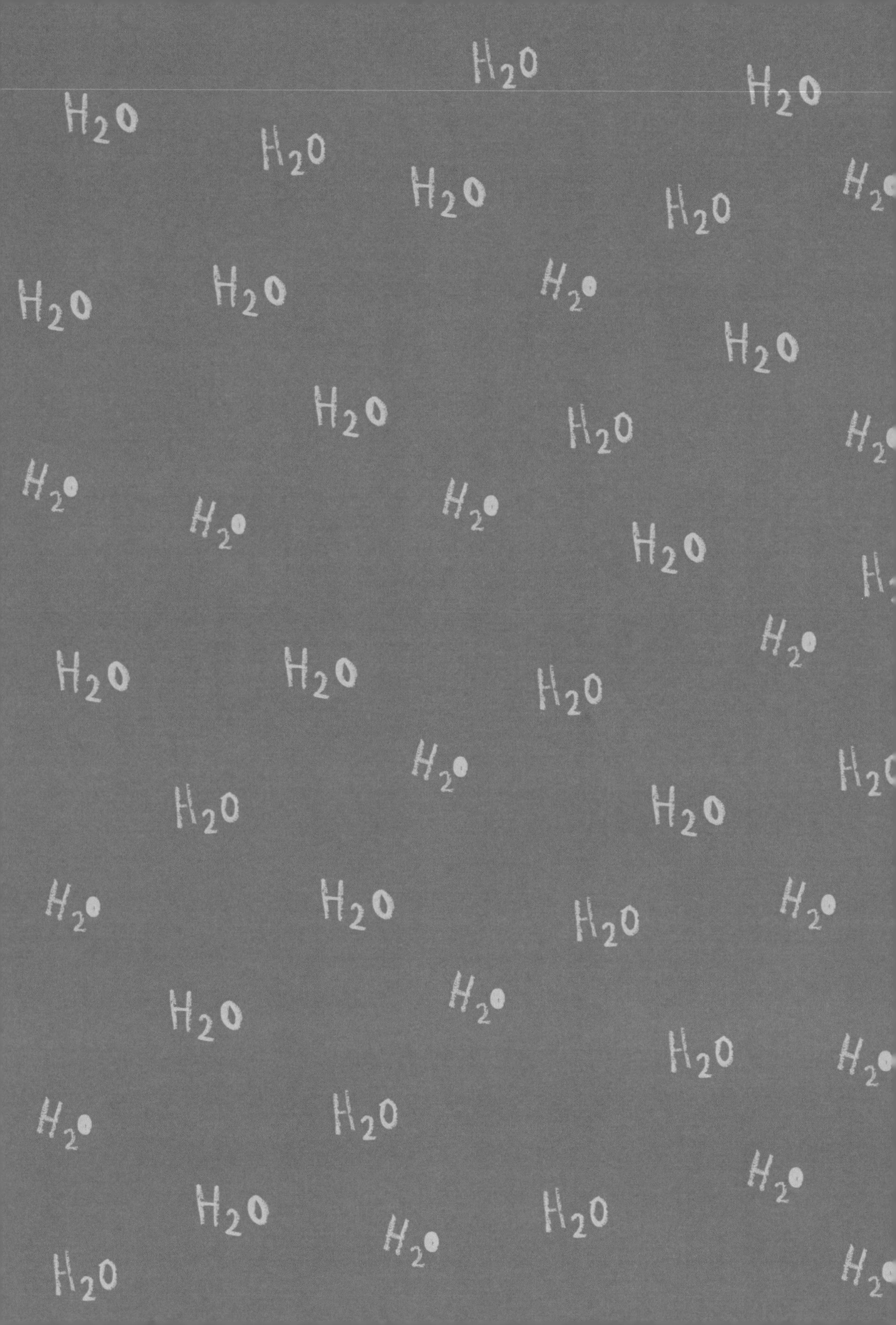

NOT ALL WATER IS THE SAME!

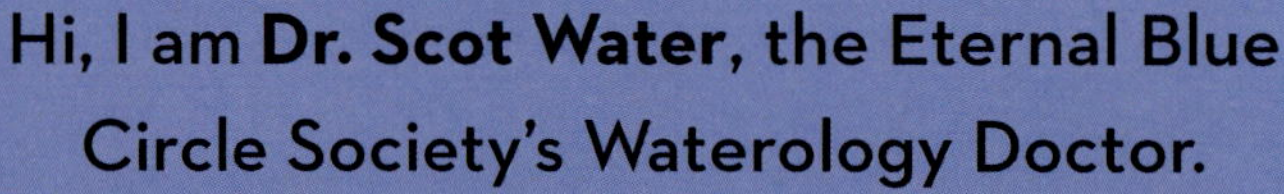

Hi, I am **Dr. Scot Water**, the Eternal Blue
Circle Society's Waterology Doctor.
To welcome you, I would like to treat you to a
nice glass of seawater... do you fancy some?
Here, cheers!
No? You don't... Well done! This means you
already know that not all water is the same!

The water we use to drink, wash
ourselves and water our plants, is
not the same as the water we dive in
during the summer at our beloved
beach.

"So, what is the difference?"

"Seawater contains salt..."
"Exactly! And it is not just table salt, the one we put in our
food, but lots of different kinds... which are called 'salts' in
Chemistry."

Seawater can't quench our thirst! Only fresh water can.

If we drink a lot of seawater, so much salt will have entered our bodies that our body will immediately try to get rid of it. In order to dissolve that much salt, our bodies will use not only the water we drank but also the fresh water stored in our cells.

Therefore, by drinking seawater, not only will we not quench our thirst but we will also waste the water our body has saved. That means we will get even thirstier...

and dehydrated!

There is another big difference:
Contrary to fresh water, salty water is...
WAY WAY WAY MORE ABUNDANT!

Let's imagine that all of Earth's water fills 100 glasses...

97 of them contain salty water!
Only 3 glasses contain fresh water!
Out of those 3 glasses, 2 are filled with glaciers, ice and snow...
So, we only have 1 left.

Out of 100 glasses of water on the planet, one glass has to be enough for drinking – for all of us... but, is it?

You will not like the answer.

NO!

It is not enough!

The amount of fresh water we have is too little.

Water may be renewed by the water cycle; it may reach the sea, vaporise, fall back onto the Earth again and change forms as it travels around in its cycle, but no one brings brand-new water to our planet.

Now that I think about it, it would be great if a space aquifer paid us a visit every day to fill our lakes...

Good idea!
I will note this down to examine the possibility further.

And as if that was not enough, **people have been multiplying** over the years; there are more and more of us. And as if that was not enough, **we use more and more water** for our needs! And as if all that was not enough, the remaining **water is often polluted** with chemicals and contaminated with microorganisms by us - by human activities, such as industry, agriculture, home sewage systems - and we can no longer use it...

As we speak there are 1 billion people that **do not have access to clean water** like we do... according to predictions, even more people on Earth may soon have trouble accessing clean water.

Just imagine what the plants and the rest of this planet's living organisms will have to go through...

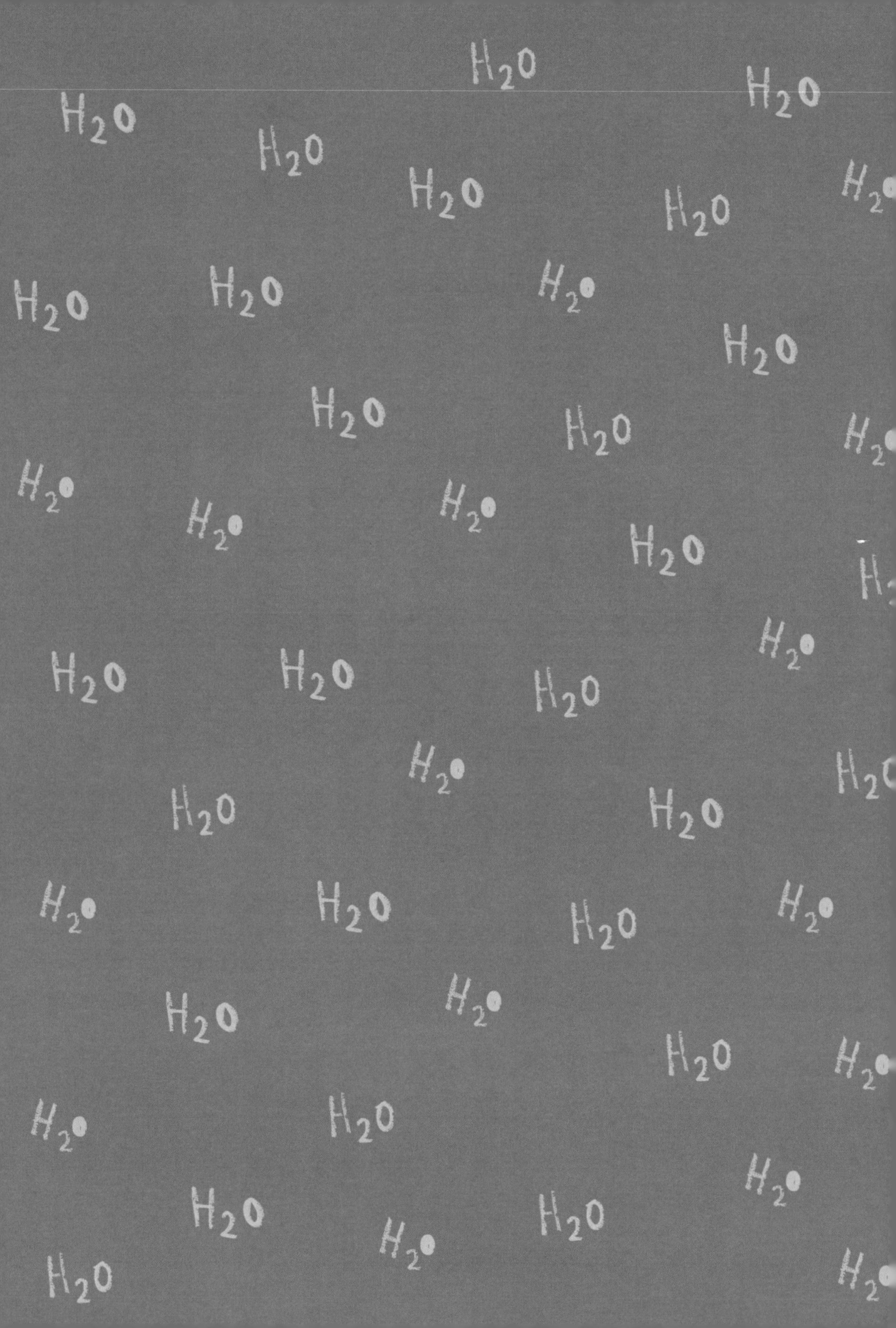

THE WATER WE THROW AWAY

Now, I will tell you a story and you have to guess how it ends:

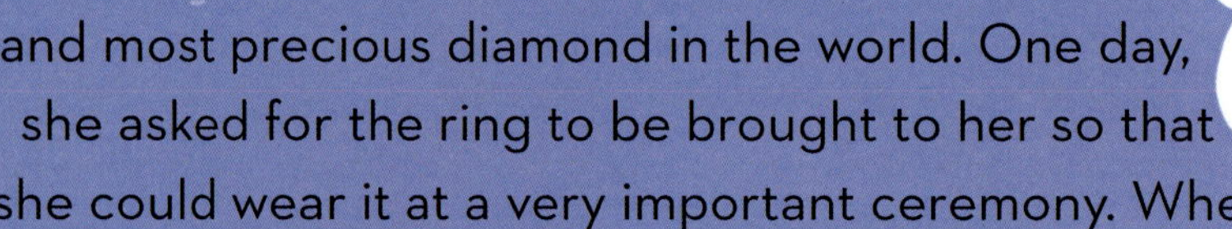

Once upon a time there was a Queen. In her safe she had a ring that carried the rarest and most precious diamond in the world. One day, she asked for the ring to be brought to her so that she could wear it at a very important ceremony. When the ceremony was over, she returned home tired, took the ring off and **what do you think she did with it?**

A) She put the ring back in the safe to wear it again in the future.

B) She threw the ring in the rubbish bin.

Here is what a water rubbish bin looks like!

Every day this water rubbish bin is filled with water from:

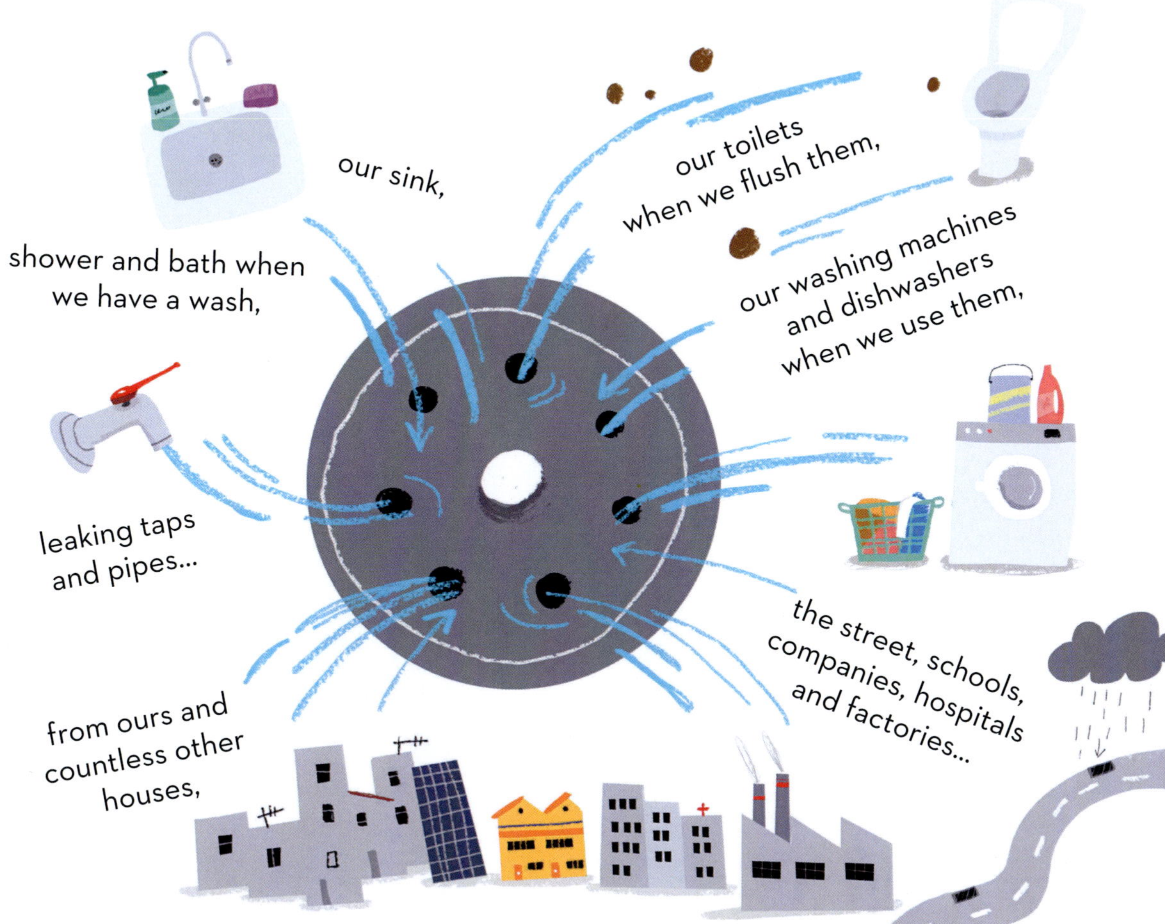

Tonnes of water end up in this water rubbish bin called the sewage system, whereas the smartest thing to do with this valuable water would be to **REUSE IT**!

But it is dirty, containing all the dirt we wash away with it; we have to clean it! This is where our Chief Cleaner, **Dr. Glow McCrystal** takes over.

Dear new members, welcome!
They say optimists see the glass half-full whereas pessimists see it half-empty! Well, a responsible Eternal Blue Circle Society member does not bother with all that... they bother with what exactly is in the glass, or the bucket, or the pipe. Because the water we throw away, used water, is not just hydrogen and oxygen like Dr. Tom Atom told you. Easy for him to say...

Whatever was on your body, the clothes, floors, dishes and cars we washed; soap, detergent, food, drink, cooking oil, colourants, sand, chemical residues, faeces and urine from our toilets, often mixed with used toilet paper, medicines we threw away down the toilet or sink, or we consumed and which passed to the toilet via our urine, bacteria from human faeces and so much more.

And, of course, we cannot use it – polluted as it is!
It would be too dangerous; this dirty water, which
we call **urban wastewater** could cause several illnesses in humans,
as well as environmental 'diseases'.
Therefore, in order to release it back to the environment,
and even further, to reuse it, we will need to clean it!

Now, that's the question:
You know how to use water to clean…

Take a glass of water and fill it with Lego or other small parts of your toys.

In another glass, use a strainer and empty all the water from the first glass into this one. See that? You just cleaned the water.

Now, try this with sand. The strainer won't help.

Now, take a clean glass, put a paper napkin in the strainer as a filter and try again. See that? Clean again!

But try dissolving a teabag in the water. Now try your filters. No use! You see, it's not always easy!

Here is where experts like us come in...
Get ready; we will travel to a water-washing machine:
one of the **Eternal Blue Circle Society's Urban Wastewater Treatment Plants...**

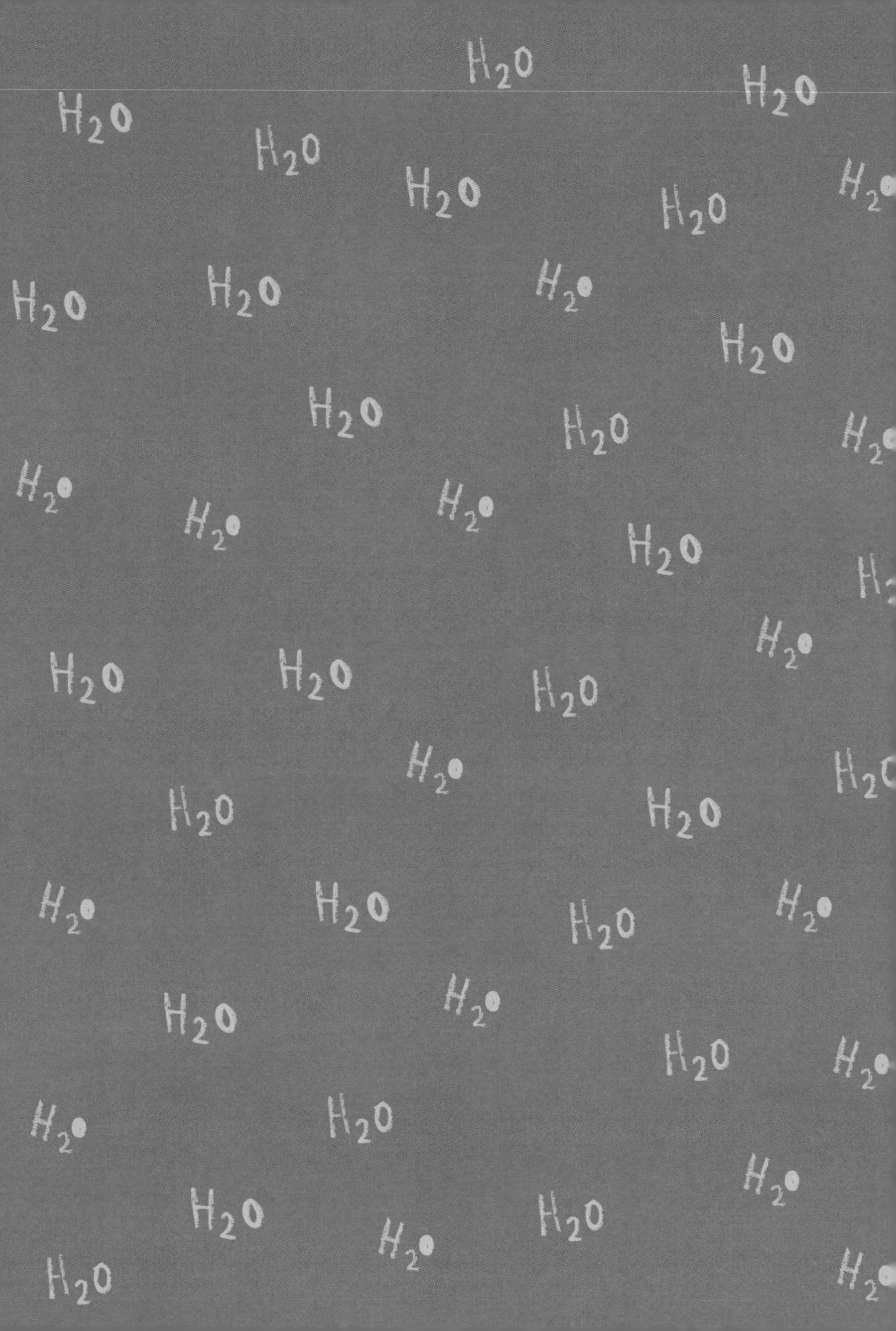

CLEANING DIRTY WATER

Welcome, I am **General Claire Clean** and I manage this Treatment Centre, where I am about to give you a tour. You must be wondering why the Centre is managed by a General. Because this is where we keep our weapons, of course! These are the weapons we use to 'wash' dirty water and fight the pollutants that are in it!

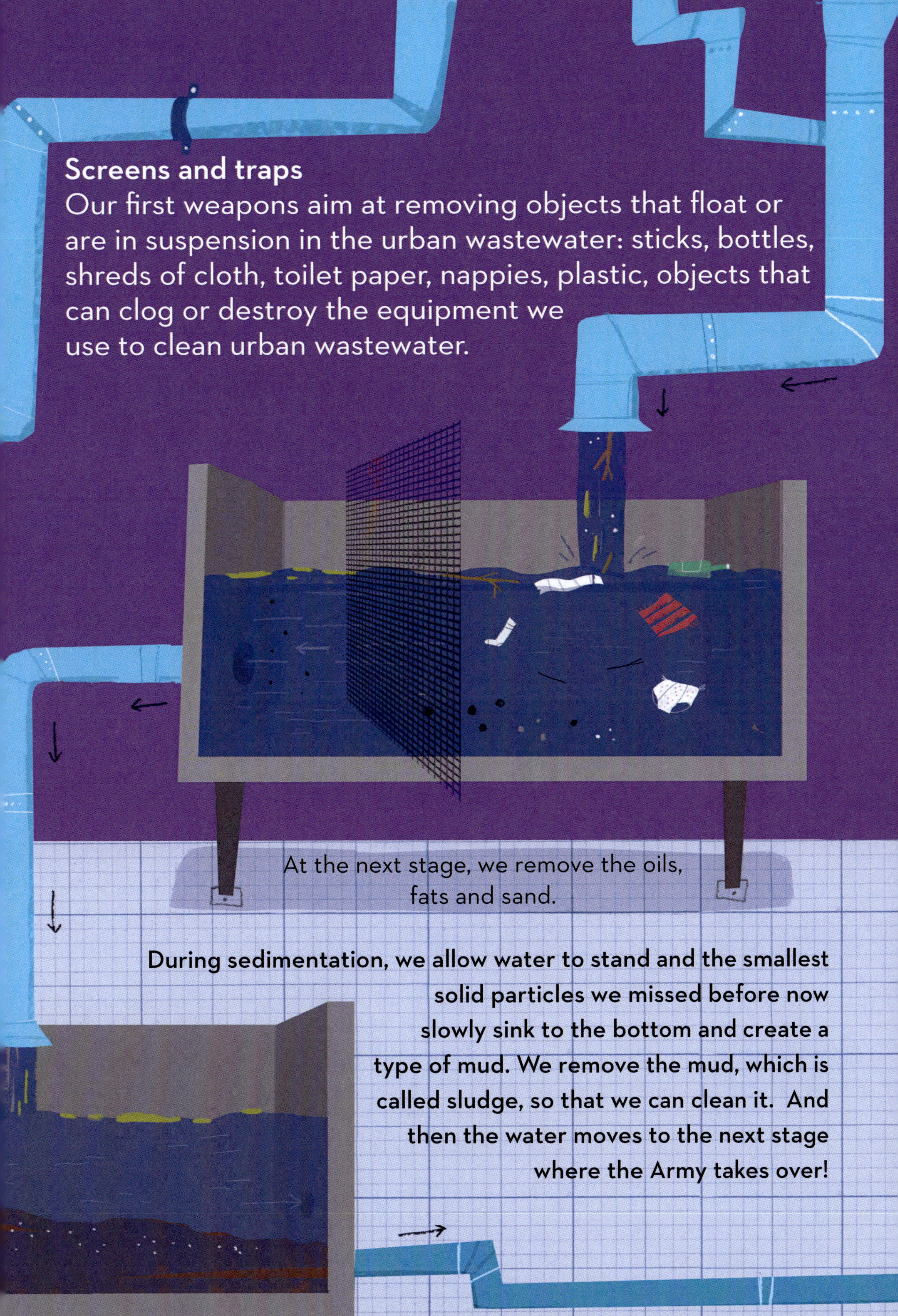

Screens and traps
Our first weapons aim at removing objects that float or
are in suspension in the urban wastewater: sticks, bottles,
shreds of cloth, toilet paper, nappies, plastic, objects that
can clog or destroy the equipment we
use to clean urban wastewater.

At the next stage, we remove the oils,
fats and sand.

During sedimentation, we allow water to stand and the smallest
solid particles we missed before now
slowly sink to the bottom and create a
type of mud. We remove the mud, which is
called sludge, so that we can clean it. And
then the water moves to the next stage
where the Army takes over!

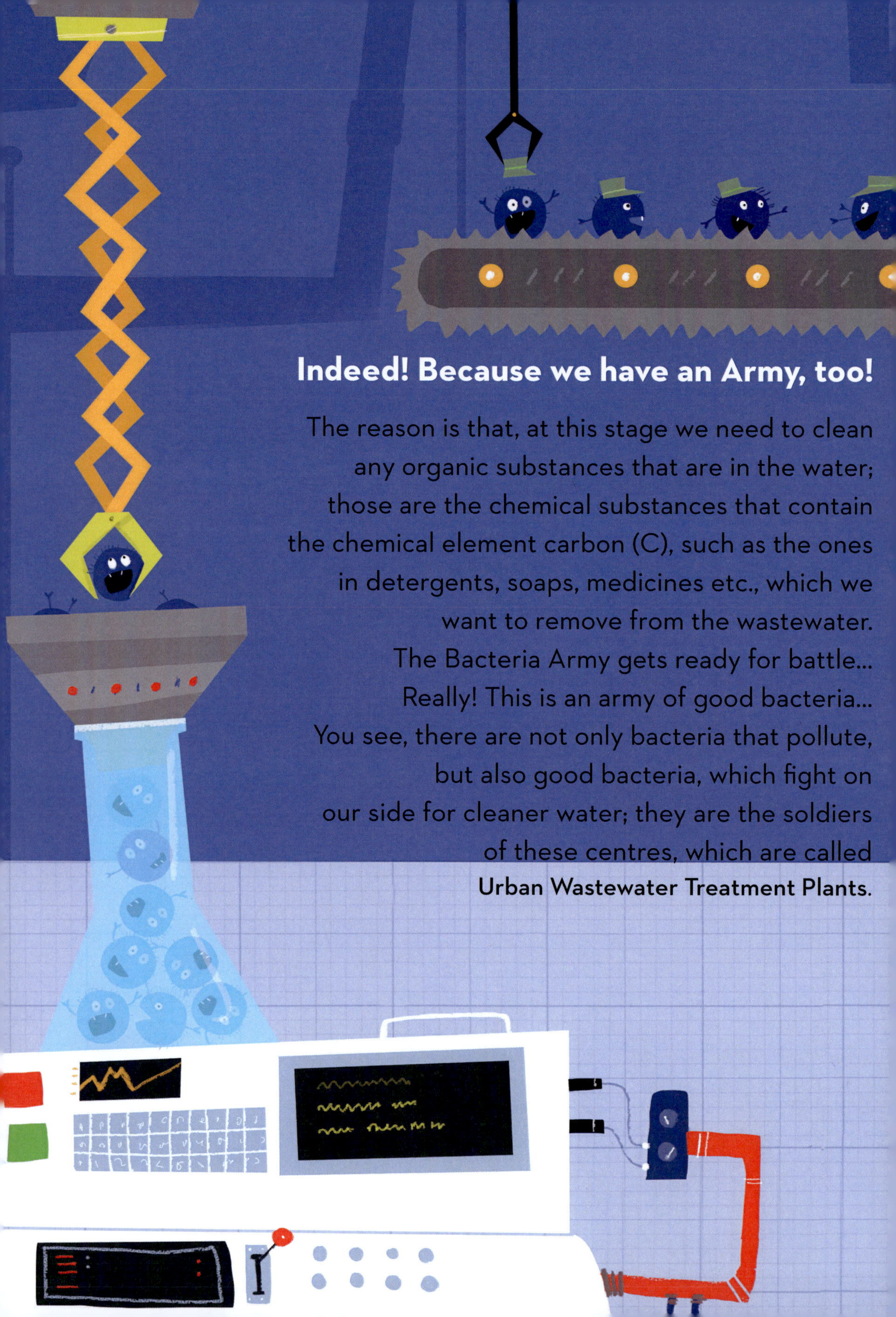

Indeed! Because we have an Army, too!

The reason is that, at this stage we need to clean
any organic substances that are in the water;
those are the chemical substances that contain
the chemical element carbon (C), such as the ones
in detergents, soaps, medicines etc., which we
want to remove from the wastewater.
The Bacteria Army gets ready for battle…
Really! This is an army of good bacteria…
You see, there are not only bacteria that pollute,
but also good bacteria, which fight on
our side for cleaner water; they are the soldiers
of these centres, which are called
Urban Wastewater Treatment Plants.

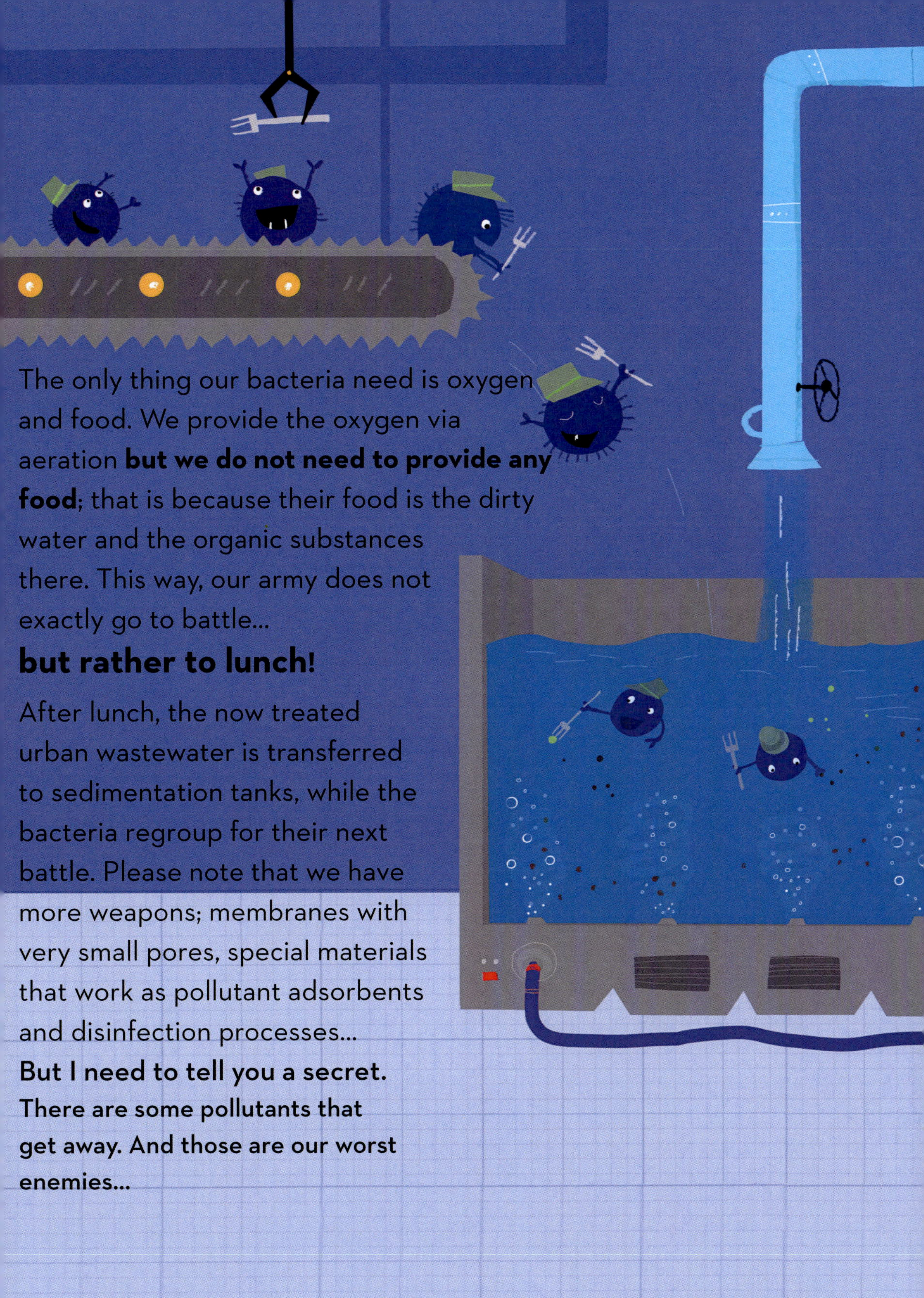

The only thing our bacteria need is oxygen and food. We provide the oxygen via aeration **but we do not need to provide any food**; that is because their food is the dirty water and the organic substances there. This way, our army does not exactly go to battle...

but rather to lunch!

After lunch, the now treated urban wastewater is transferred to sedimentation tanks, while the bacteria regroup for their next battle. Please note that we have more weapons; membranes with very small pores, special materials that work as pollutant adsorbents and disinfection processes...

But I need to tell you a secret. There are some pollutants that get away. And those are our worst enemies...

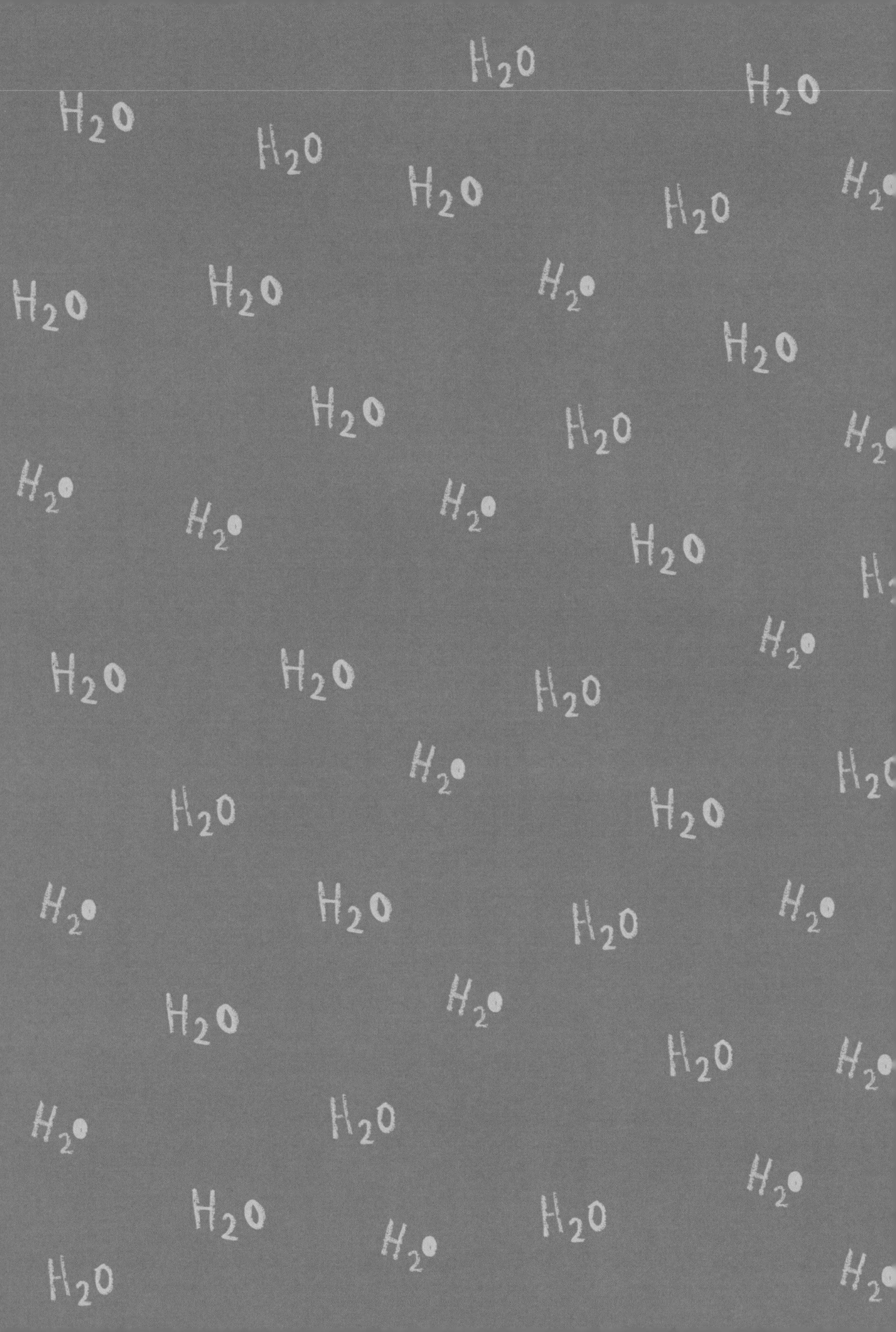

PERSISTENT POLLUTANTS

They are our worst enemies because they manage to run away with the treated urban wastewater that returns to the environment (e.g. rivers), or the treated wastewater we reuse (e.g. for watering).

Pollutants from:

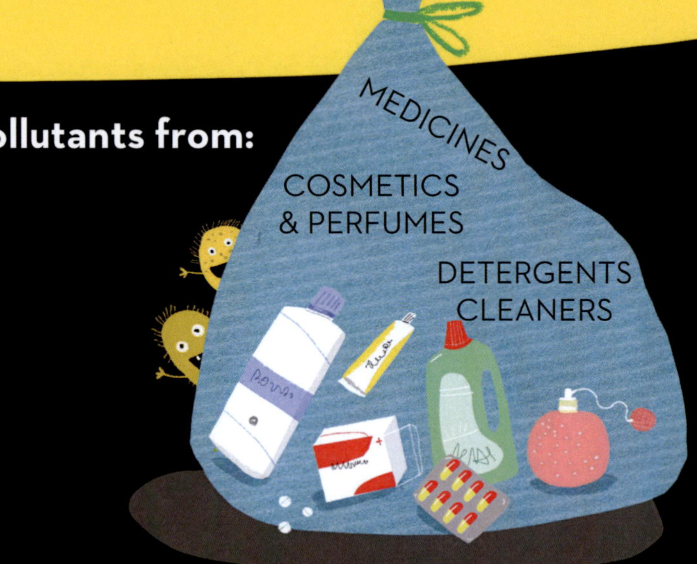

These are important and necessary substances for us when used but, as soon as they enter the water, they become POLLUTANTS. Pollutants that only exist in very small quantities in urban wastewater, are smart, know how to hide, twist and turn, change and persist. These are pollutants that attract the interest of us scientists, more and more; that is why we call them 'CONTAMINANTS OF EMERGING CONCERN'.

Dr. André Antib will tell you more...

An infection occurs when microorganisms get in our body and start to multiply so fast that our body can no longer fight them. Then we go to the doctor and they examine us, combining several clues, using their knowledge and experience, to decide whether to give us antibiotics.

An antibiotic is an organic substance that can kill the bacteria that trouble our body; it is a POISON. But a poison only to bacteria - not to us! And that is why it is a great discovery.

Those of you who have taken antibiotics must have noticed that the doctor's instructions are always very STRICT: this many times per day, this much medicine, for this many days. That is because the doctor knows bacteria are clever microorganisms. If the antibiotic is not used correctly and fails to kill them, the bacteria will learn the antibiotic's secrets and will know how to avoid it. This means they will become resistant, get used to it and find ways to fight it and survive.

This will not happen if we stick to our doctor's instructions; our illness will go away and we will get well soon.

However, the bacteria that have learnt our antibiotic's secret, where are they going to end up via our toilets and sewage systems?

Think about it...

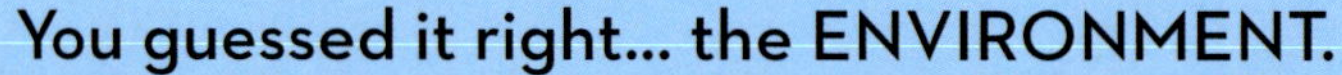

You guessed it right... the ENVIRONMENT.
They will go through all narrow and wide pipes, reach the treatment plant and some of them will get past all of our weapons undisturbed, and when the treated urban wastewater is discharged (for example in rivers) or when it is used for different purposes, (such as watering our city's green spaces) they will manage to escape...
INTO NATURE!

There, they will multiply and meet more bacteria...
That is how our antibiotics' secrets will spread and more bacteria may become resistant...
As a result, next time doctors have to fight against one of them, they may need to come up with a new antibiotic.

And it is not just bacteria. There are also those substances we call '**contaminants of emerging concern**', which escape from urban wastewater treatment plants and which, even in small quantities, can affect animals and plants, change their habits, their forms, and lives in ways we had never intended when we created those substances.

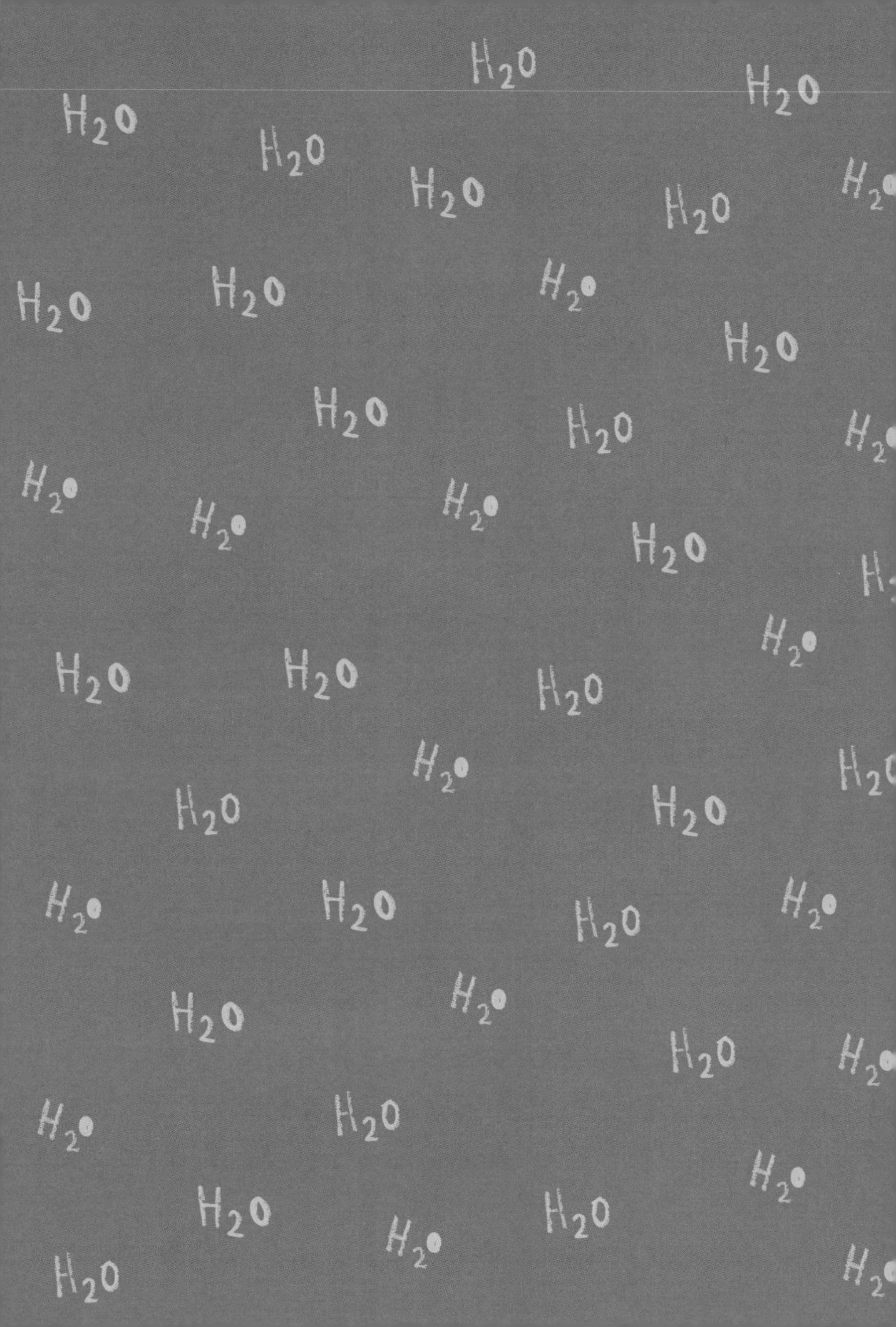

Secret

CHAPTER

8

OUR MISSION (AND YOURS)

This is our mission. And yours.

The Eternal Blue Circle Society's mission is to carefully study contaminants of emerging concern, their effects on nature, and to evaluate and document them in order to design new secret weapons which capture or destroy them.

We must find ways to control pollution.

For years now, scientists from all over the world have been coming together to become the protectors of water, the invaluable substance which maintains us and all life on our planet.

Our tricks may be kept secret – so that pollutants do not know about our new tricks – but the importance of water pollution is no secret; it has to be known by...

EVERYBODY!

Everybody needs to:

▶ Know how valuable and rare and precious water is, so that they look after it.

▶ Learn not to waste it.

▶ Learn to pollute it as little as possible.

▶ Return unused or expired medicines to their pharmacist or other appropriate collection centres.

And this is something that you, our new members, can help us achieve from now on!

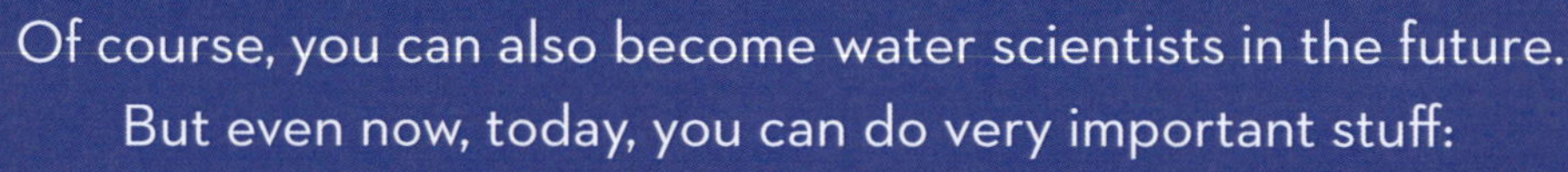

Of course, you can also become water scientists in the future.
But even now, today, you can do very important stuff:

- Make your parents, friends and classmates aware.

- Organise a classroom presentation about water and how to protect it.

- Prepare a poster about water.

- Shoot a documentary about water and make sure it is seen by as many people as possible.

- Start a school newspaper about water.

- Start your own show about water.

- Organise a party about water.

★ (*all of the above should be done with extra discretion, as the persistent pollutants' spies live among us)

And send us a report of what you did at:
bluecirclesocietykids@ucy.ac.cy
so that we can upload it on our website:
http://nireas-iwrc.org/
and send you and your classmates your certificate!

Dear reader,

If you have made it to this point, now you know what you need to do to become an Eternal Blue Circle Society member and a protector of water on Earth.

You are now one of us.

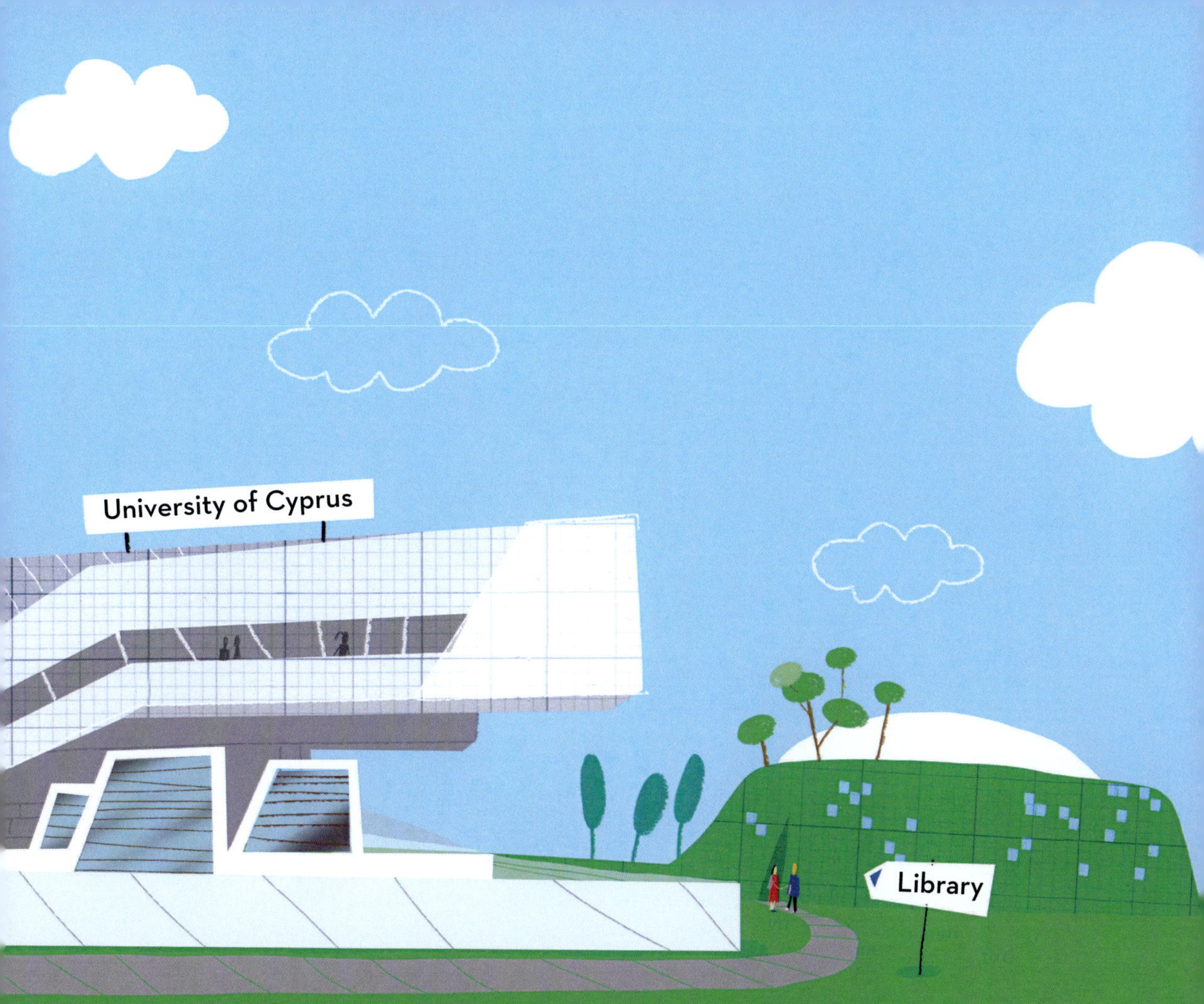

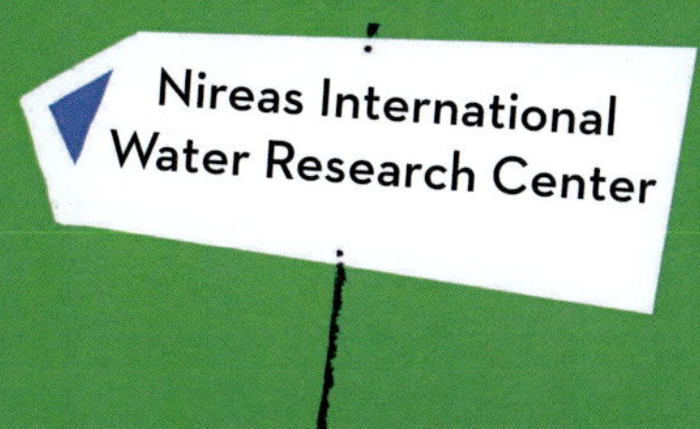

Do not forget to defend water wherever you are and always wear something blue so that we can recognise you:
A blue superhero mask, a huge blue ribbon, a big blue cape or a very tall blue hat.
Or anything else of your choice in blue, in case the above sounds a bit too much to you.

Despo Fatta-Kassinos is from Famagusta, Cyprus. After the Turkish invasion her family moved to Larnaca, where she grew up. She graduated from St George's High School and went on to study Chemical Engineering at the National Technical University of Athens (NTUA). After completing her MSc in Greece and Italy, she was awarded her PhD from NTUA.

She is currently an Associate Professor in the Department of Civil and Environmental Engineering, and the Director of the Nireas-International Water Research Center at the University of Cyprus.

Her research focuses on understanding the fate and behavior of contaminants of emerging concern during advanced water and wastewater treatment and within the environment. Specifically, the activities of her research group include the identification of xenobiotic compounds in environmental samples, the assessment of their environmental impact, the development of new technological processes for their degradation and removal from aqueous media, and the implications of wastewater reuse applications for the environment.

She is Editor of the Journal of Environmental Chemical Engineering, Elsevier, and the leader of Working Group 5 'Wastewater reuse' of the NORMAN Association. She was the Chair of the NEREUS COST Action ES1403, which aimed to provide answers regarding the actual effects of reusing treated urban wastewater effluents in the environment. She is the Coordinator of the European research project ANSWER (H2020-MSCA-ITN-2015/675530) within the framework of the Marie Sklodowska-Curie: Innovative Training Networks (ITN) Action, which is training a new generation of creative early-stage researchers in the interdisciplinary approaches required to meet the major challenges facing the field of wastewater treatment and reuse, and to be able to convert knowledge and ideas into products and services for economic and social benefit. She served as the Chair of the Scientific and Technological Advisory Board of the European Joint Programming Initiative 'Water Challenges for a Changing World' (2015-2019). She has published over 145 scientific papers, edited various environmental books, and has participated in more than 170 scientific conferences and over 50 research programmes.

Irene Michael-Kordatou is a Chemical Engineer (NTUA, 2008) and holds a PhD in Environmental Engineering from the University of Cyprus (2012). She has been working as a senior researcher at Nireas-International Water Research Center since May 2012.

Lida Ioannou-Ttofa has a degree in Chemical Engineering from the Aristotle University of Thessaloniki (2008), an MSc (2010) and a PhD in Environmental Engineering from the University of Cyprus (2013). She has been working as a senior researcher at Nireas-International Water Research Center since November 2013.

Costas Michael holds a PhD in Chemistry from the University of Athens and is a former Director of the State General Laboratory of Cyprus. He is a member of the Nireas-International Water Research Center Board of Directors.

Dr. Despo Fatta-Kassinos, as the Chair of the NEREUS COST Action ES1403 and the Coordinator of the ANSWER project (H2020-MSCA-ITN-2015/675530), would like to acknowledge the valuable funding received for scientific work in the field of wastewater reuse and related emerging challenges. This book was inspired by this work.

NEREUS, COST Action ES1403:

New and emerging challenges and opportunities in wastewater reuse

COST is supported by the EU Framework Programme Horizon 2020

COST (European Cooperation in Science and Technology, http://www.cost.eu) is a funding agency for research and innovation networks. COST Actions help connect research initiatives across Europe and enable scientists to grow their ideas by sharing them with their peers. This boosts their research, career and innovation.

ANSWER, H2020-MSCA-ITN-2015/675530:

Antibiotics and mobile resistance elements in wastewater reuse applications: risks and innovative solutions

This project has received funding from the European Union's Horizon 2020 research and innovation programme under the Marie Skłodowska-Curie grant agreement No 675530.

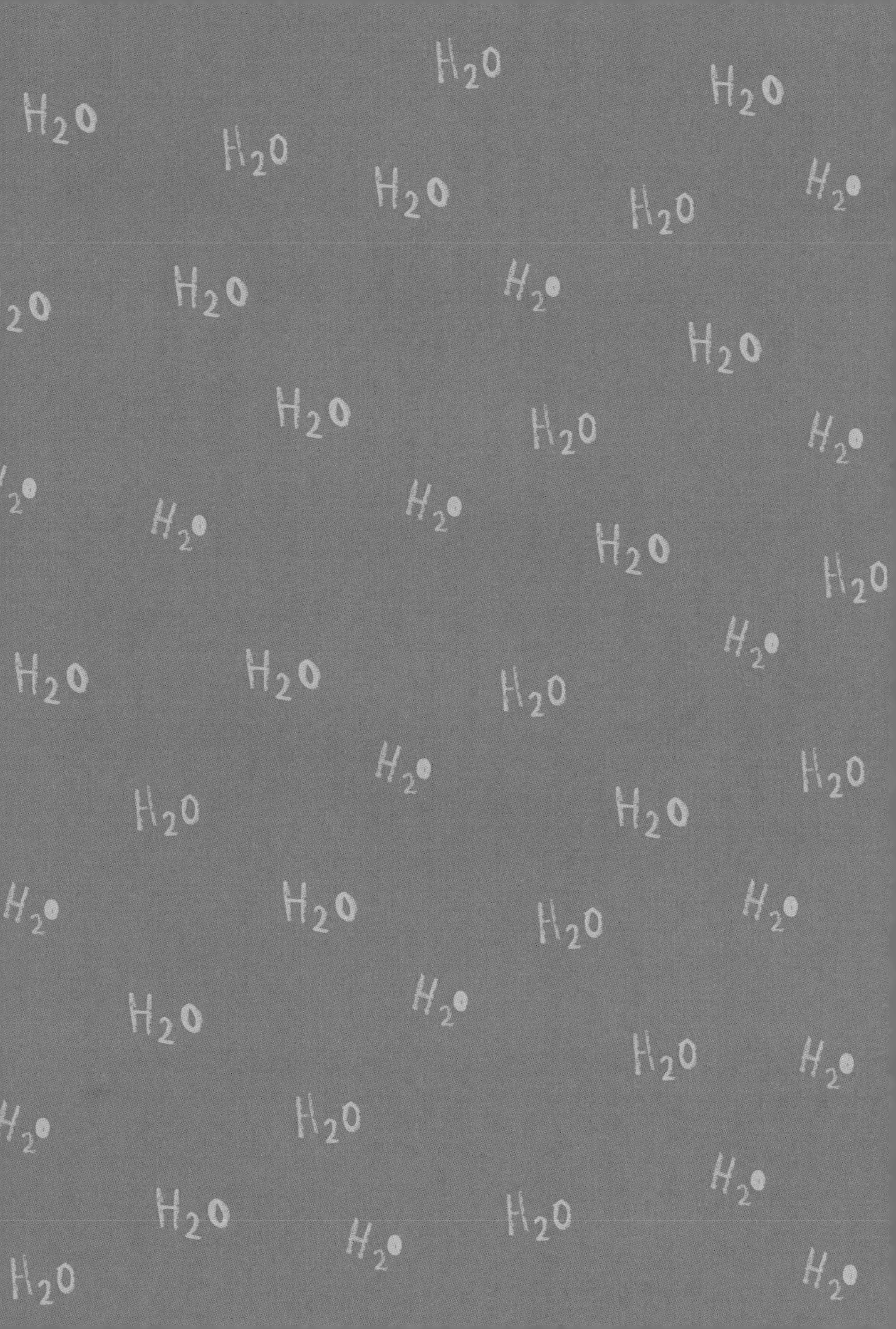

IWA Publishing's authorised EU representative for General Product Safety Regulations is Diane D'Arras, 15 rue Duret, 75116 Paris, France, e-mail: safety@iwap.co.uk.

Printed and bound by CPI Group (UK) Ltd, Croydon, CR0 4YY

12/05/2026

02108885-0001